Hamburg Chickens
From The Book of Poultry

by Lewis Wright

with an introduction by Jackson Chambers

This work contains material that was originally published in 1891.

This publication is within the Public Domain.

This edition is reprinted for educational purposes
and in accordance with all applicable Federal Laws.

Introduction Copyright 2017 by Jackson Chambers

Self Reliance Books

Get more historic titles on animal and stock breeding, gardening and old fashioned skills by visiting us at:

http://selfreliancebooks.blogspot.com/

Introduction

I am pleased to present yet another title in the "Chicken Breeds" series.

The work is in the Public Domain and is re-printed here in accordance with Federal Laws.

Though this work is a century old it contains much information on poultry that is still pertinent today.

As with all reprinted books of this age that are intended to perfectly reproduce the original edition, considerable pains and effort had to be undertaken to correct fading and sometimes outright damage to existing proofs of this title. At times, this task is quite monumental, requiring an almost total "rebuilding" of some pages from digital proofs of multiple copies. Despite this, imperfections still sometimes exist in the final proof and may detract from the visual appearance of the text.

I hope you enjoy reading this book as much as I enjoyed making it available to readers again.

Jackson Chambers

HAMBURGHS.

THE fowls known at the present day under the general name of Hamburghs, and which present the common characteristics of rather small size, slender clean legs, neat rose combs, moderate-sized white ear-lobes, light but sweeping and graceful outlines, and the absence of the incubating instinct, had certainly two distinct origins. There is very little doubt that the two varieties known as Pencilled Hamburghs were really imported from Holland, having been, for years previously to the present name, well known under the title of Dutch Everyday Layers or Everlasting Layers, and having been largely imported both at a comparatively recent date, and on many former occasions of which there is historical evidence. The Spangled and Black varieties, on the contrary, are as evidently a native English breed of considerable antiquity, having been kept and shown at village exhibitions beyond the memory of man; and the Spangled breeds, besides the difference in marking, presenting differences of shape, being larger, plumper, and somewhat coarser in make, besides a greater width across the skull. These latter varieties were always known under the name of the Lancashire Mooneys and Yorkshire Pheasant fowls, while the Blacks were called Black Pheasant fowls; until almost immediately after the establishment of the Great Birmingham Show the authorities there—chiefly, we believe, owing to the influence of the Rev. E. S. Dixon—grouped all under the general name of Hamburghs, and the paramount authority of that great gathering was sufficient to make the new nomenclature general. We confess we cannot see the evils some have pedantically professed to find in this; the new name is not, so far as we have found, at all misunderstood in a geographical sense, and it answers at least the useful purpose of forming into one compact group fowls which in their main characteristics are very similar.

We are not sure but that the argument may be carried further, and that the old Birmingham authorities whose work has been so much villified in some quarters may not have been guided by a sounder instinct than at first sight appears. Fully admitting that for a long period the Spangled and Pencilled races have been thoroughly distinct, and can be traced back only to distinct origins, it is still impossible to stand before the pens at a good show and compare the Gold-pencilled with the Gold-spangled, and similarly the two Silver classes, noting thus practically the striking analogy not only in heads, deaf-ears, size, and shape, but in the common property of a distinct lustrous black marking on one or the other ground-colour, and remembering also the equal similarity as regards absence of any incubating instinct, without the question almost irresistibly arising of whether there was not some still more ancient common progenitor of both; and this question is by no means dismissed by the difference in the shape and character of the present markings. Once started, the arguments in favour of such a hypothesis will be found very strong indeed; for not only, as we shall see further on, are Spangled chickens frequently *pencilled* in their first feathers, but later in life the black spangles or moons are frequently surmounted by a light tip beyond them, thus again approaching to the pencilled character; while conversely it will be seen that if Pencilled birds be bred too dark, the last bar has a strong tendency to become much too wide, thus approaching a *spangled* character. Perhaps, indeed, the evidence of these facts is about as strong as could be possibly anticipated at such a late date in favour of this view, if we consider the utter want of interest in poultry, and consequent absence of observations at any distance back; and the same reasons make it exceedingly difficult even to conjecture what the common progenitor—if there was one—may have been. Bearing in mind, however, that old Aldrovandus calls a fowl which bears a very plain general resemblance to the Silver-Pencilled breed *Gallina Turcica*, or the Turkish Fowl, it is rather interesting to notice how this very name alone suggests an Eastern and therefore probably more ancient origin than either the Dutch or English we have been considering; and it may be that here we have either the parent or a near descendant from the parent stock of each. These are, however, mere speculations; they may be legitimate, or they may not, and we mention them chiefly as another instance of the fascinating problems that constantly present themselves for solution to the poultry-fancier of a philosophic or inquiring turn of mind.

In proceeding to the consideration of the different varieties of Hamburghs, we have the greatest pleasure in being able to give the views and experience of Mr. Henry Beldon, who when living at Goitstock, Bingley, Yorkshire, was never approached in his general and continuous success as an exhibitor of the Hamburgh varieties. It has long been known that the rules of breeding followed by him and others at the present day differed considerably from those formerly pursued and recommended by fanciers of good standing; but no account of the present methods was ever published before those written by him for this work. All breeders will therefore feel indebted to him for them; while naturalists will feel interested in the light thrown upon various questions by the amalgamation of races which only a very few years back were pronounced by high authorities incapable of such real union, though crossed *as* separate varieties for the production of show birds.

"Hamburghs," says Mr. Beldon, "are without doubt the most beautiful breed of poultry we possess, as well as one of the most useful. In their different varieties they give plenty of scope to the fancier to indulge his tastes; but whether he choose Pencilled or Spangled, Gold, Silver, or Black, all are without doubt elegant and beautiful. The dweller in the country will generally prefer the Silver, while the citizen will take the Golden or the Black; but all of them, in their matchless variety of marking and colour, will delight the eye with the utmost degree which is perhaps possible of beauty in fowls.

"Hamburghs, however, require free range. They are of little or no use penned up, in which state they pine and mope for liberty, that bright cheerfulness which is common to them disappears, and from being the happiest they become the most wretched of birds. If your convenience will not allow you to give them a grass-run of moderate size, my advice is to keep some other kind; but if you have a good run no class of poultry will pay so well. They are small eaters and wonderful egg-producers, a single hen laying in a twelvemonth, under favourable circumstances, from 200 to 220 eggs. They are also capital foragers, and when in health will always be seen at work, especially in the early morning, rummaging the pastures. Their quick eye at once espies their prey, and woe to the poor worm that happens on that particular morning to have got up a little too early; its early hours are suddenly put a stop to, and in this case the early riser finds it is not always well to be up *too* soon. Another good quality is that they are generally non-sitters, and there is not so much trouble with them as with the sitting varieties, though there are exceptions to the rule. I know that some people look upon any Hamburgh sitting as an evidence of taint in the blood, but I am quite sure this is an erroneous notion; for the best Silver-spangled hen I ever possessed (in fact, the very same bird Mr. Teebay mentions in another work as one of the best he ever saw, and as moulting without change up to seven years old), without doubt the most successful winner of cups and prizes during the present generation, wished to sit. In this case, however, the fit only came upon her in old age, being no less than eight years old. She clucked and sat steadily upon any eggs she could get at, but I thought she was too old to take the charge of a family with all its cares and troubles, and therefore checked her, though with considerable difficulty. The poor old lady died in the winter of the same year (1870), and I only mention her as showing that the very best birds will sit occasionally.

"As a rule Hamburghs are a healthy breed, and for the farmer I think they are the fowl of fowls. On a good homestead they will almost keep themselves, and if well attended to will pay as well as any other part of his stock. The chickens, too, are easy to rear. Of course, they will not rear themselves, but with moderate care no difficulty will be found in getting them to maturity. They need good coops, which should be placed on a nice grass-run, as far as convenient from the old birds, as I always find chickens do much better *quite away from adults*; otherwise the old fowls pick up the food the chickens should have, not only robbing them, but getting far too much for their own welfare. They should be fed often, giving only a little at a time, just what they will eat. In the first five or six weeks, I should advise feeding every two or three hours, after that less often will do; but the better the chicks are looked after the finer they will be, and there is nothing lost by a little extra care. The best staple food I have found to be oatmeal and thirds mixed, and made up into a stiff crumbly mass. As they get older I mix more of the thirds and put less oatmeal, and by degrees give a little wheat, but soft food should be the chief of the diet. The great thing is to feed often, beginning early and leaving off late. The coops I prefer are simply made of wood, about two feet square, with sloping roof and sliding front, to admit of letting out the chickens without the hen if you think proper, but with a movable bar to let out the hen also. The coop should be without bottom, so that you can change on to fresh ground as the other gets tainted, and if possible I change on to fresh ground daily. The chickens reach maturity early if well cared for and not stopped in their growth. I have often had pullets laying at five months old, especially of the Pencilled varieties; the Spangled do not generally lay quite so early.

SILVER-SPANGLED HAMBURGHS.—"The Silver-spangled Hamburgh, or Silver Pheasant as it is commonly called in Yorkshire, is a breed that has for generations been known in this country, and for its cultivation to the present state of perfection owes everything to the counties of Lancashire and Yorkshire. In Lancashire this variety had been brought to a very high standard of excellence years before ever poultry-shows were thought of, and as regards feather, all our modern skill and careful breeding has been unable to improve upon the old breed; indeed, I don't think it would be possible to improve it, for some of the old Mooneys, as they were called, were absolute perfection in this point of feather; the spangling, so large, round, and rich in colour, was really something to be wondered at, and shows a skill and enthusiasm in breeding which, in the absence of public shows in those days, has about it something of the marvellous.

"This careful and extreme breeding for feather in the old Lancashire Mooney fowls it was, in my opinion, which resulted in producing hen-feathered cocks—that is, cocks feathered similarly to the hens, with spangling on back, sides, neck, &c., and with a square or hen-tail (Fig. 82). Be this as it may, it was to this variety at the beginning of the poultry-showing era, a good many years since, that all the prizes were given; but after they had enjoyed a year or two's popularity, the judges at Birmingham all at once announced that this hen-feathering of the cocks was not the 'correct thing,' and also stated that such birds were unprolific. In this latter charge there was some truth, as many of these cocks will not breed, though some others are prolific enough; but as a result the hen-tailed cocks were thrown out, and their reign as show birds was over, though they are still kept by a few ardent fanciers for breeding purposes solely.

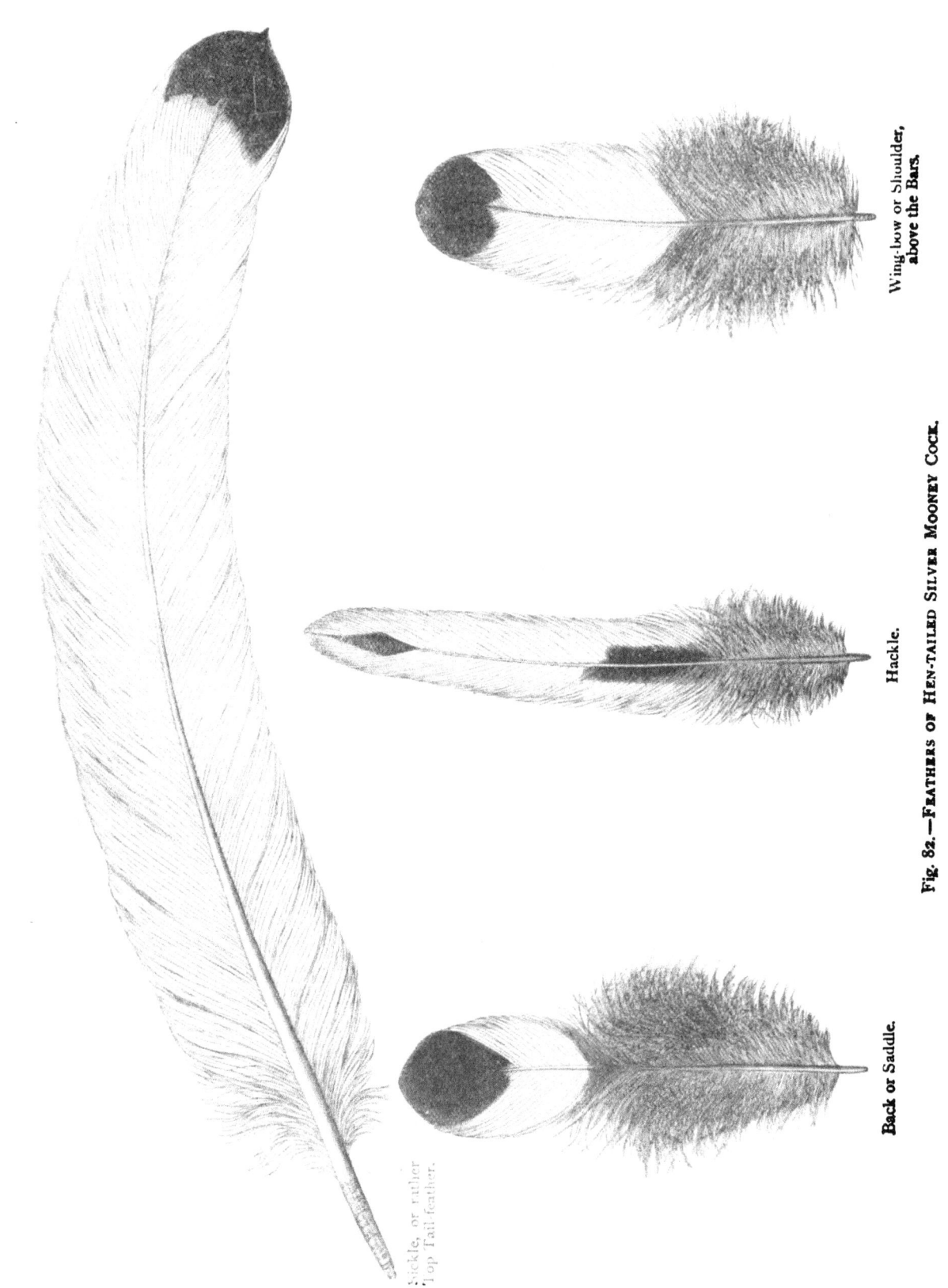

Fig. 82.—Feathers of Hen-tailed Silver Mooney Cock.

Sickle, or rather Top Tail feather. Hackle. Wing-bow or Shoulder, above the Bars. Back or Saddle.

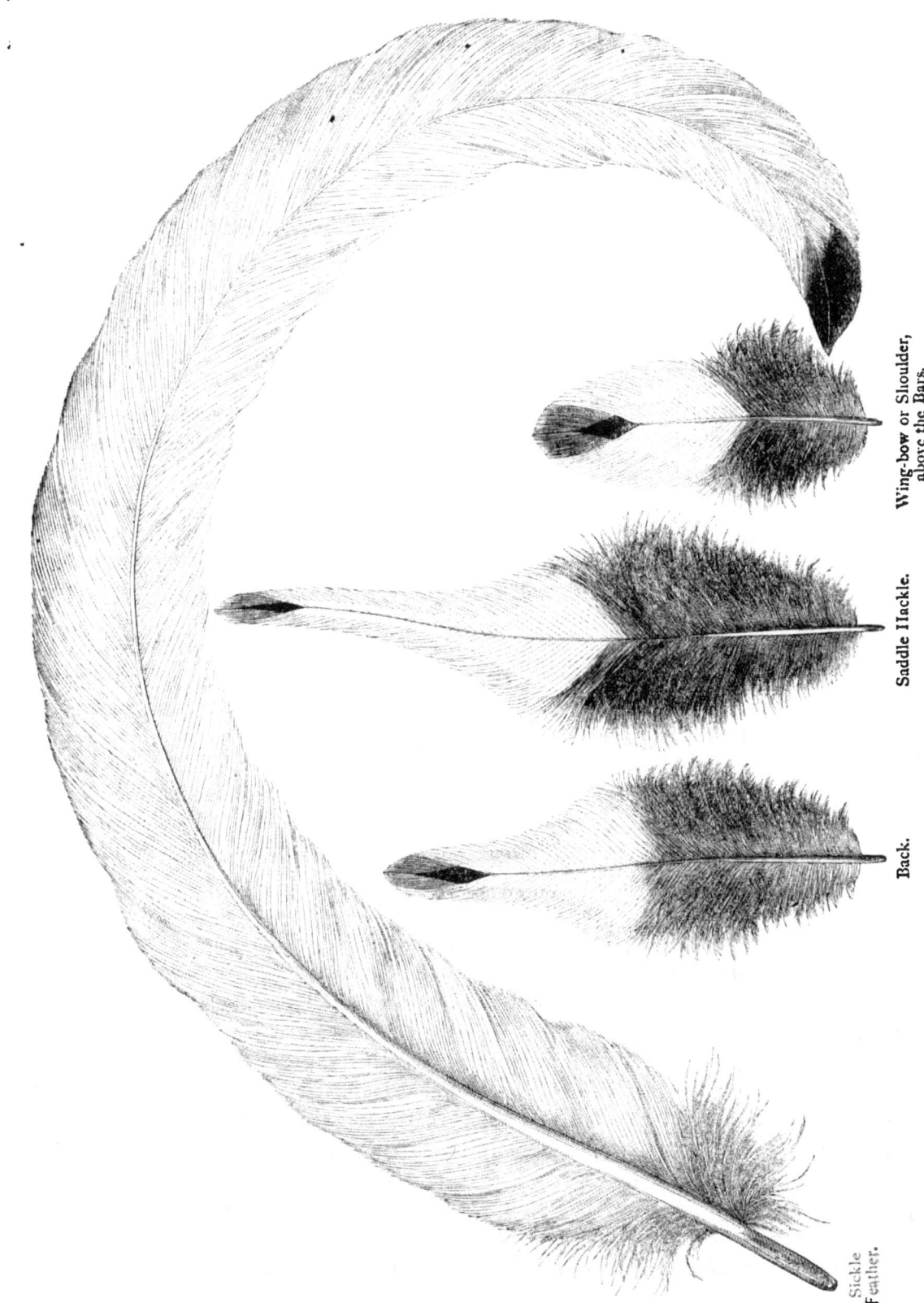

Fig. 83.—Feathers of Full-plumaged Silver-spangled Cock.

"In Yorkshire, on the other hand, we possessed another Silver-spangled breed which had the desired cock-feathering, the cock being indeed a fine full-plumaged bird; but the colour and size of the spangling was much inferior to the Lancashire variety. I feel morally certain myself that this was the breed the Lancashire people had to commence with, and which was bred up to such perfection by them as stated above, losing in ear-lobe and developing the hen-tail through continued breeding for feather only. This cock-feathered, or Yorkshire Pheasant breed as it was called, lacked not only the size, but the roundness and glossy greenness in the spangles of the Lancashire variety, but were decidedly smarter in appearance, and possessed whiter ear-lobes. The hen-feathered Lancashire Mooney breed has reddish ear-lobes, and even the Yorkshire Pheasants had not much to boast of in this respect; still they were whitish, and a few years' careful breeding soon brought this point to perfection. The Yorkshire cocks had, moreover, nice clear tails, while the hen-feathered cocks often had smutty tails; but, on the other hand, they much lacked colour on the back and wings.

"At first these Yorkshire cocks were shown alongside the Lancashire Mooney hens, and of course to breed prize-winners the two varieties had to be kept and bred separately. Each variety, however, possessed great defects: in the hen-feathered Mooney the combs were coarse and the ear-lobes red; while in the Yorkshire cocks the back, saddle-hackles, and shoulders were white. The two were therefore bred together, at first chiefly for the production of better cockerels; but by degrees an amalgamation of the two breeds was brought about, and by careful and judicious crossing a bird was at last produced that contained all the required characteristics. Some noted fanciers yet breed from two sets for cockerels and pullets, taking for pullets the old hen-feathered breed; but when this is done great care must be taken that the strain is pure hen-feathered on both sides. The disadvantage of this plan is, that the surplus cocks from this strain are worth absolutely nothing for exhibition, and very frequently will not breed [*i.e.*, are unprolific], so of course such surplus birds have to be killed off, or sold for breeding purposes only. For cock-breeding, of course, such fanciers choose birds likely to produce the required points; seldom now using the pure Yorkshire, but selecting a bird with fine smart comb, good ear-lobes, good bars, well spangled breast, and as clear a tail as possible, with of course good back and saddle spangling (never found in the Yorkshire pure), and putting him with hens well spangled throughout, and having good combs and ear-lobes. It will, no doubt, be thought by many to be but a clumsy way of breeding, to have to breed from what are really two sorts; and in fact it is now not really necessary to do so, as the cocks now suited for the most successful competition are also such as breed the best pullets; the chief requirements necessary to success in the pen being now that the bird possess as much marking as possible without being hen-feathered, which is also what we want for pullet-breeding. This has taken much time and patience to effect, but it is unquestionably a very great gain.

"I therefore, taking the Silver-spangled breed as it has been formed by the skill of fanciers, and now actually exists, advise the beginner to proceed as follows. Let him get from some well-known breeder such a bird as I have just described—that is, possessing good comb and ear-lobes with as much spangling on back and saddle as possible, good bars, and clear tail, but not hen-feathered; in fact, a good deep-coloured show cock; and simply put him to the very best hens he can get, avoiding carefully any great faults on either side, such as a coarse comb or smudgy markings, and then try what he can do. If he thus produces a fair proportion of good chickens, let him stick to this set as long as they will breed, for it is not every lot that hits well. If the produce is not to his mind let him change the cock, getting one from some other strain, and so on till he gets what he desires; for the different strains and the two breeds I have described are now so mingled that it is difficult to proceed in any other way. In all varieties of fowls there are found some strains which produce better cockerels than pullets, and *vice versa*. Of course, it is so in Hamburghs also; and if the fancier has sufficient room no doubt he will find it to his advantage to breed from two sets. For instance, if he finds a pen breed capital cocks but only middling pullets, it will be better to keep this set of birds as they are, it being far better, and perhaps harder, to breed really good birds of even one sex, than middling birds of both. It is therefore well worth a little patience, and when you have once got a set of birds together that produces first-class chickens, then stick to that set. All experience will confirm this; and with all my own, taking Hamburghs as they are at the present day, I do not know that I can give any other rule, which is that by which I breed my own fowls. We never now breed from the Yorkshire Pheasant if we can possibly help it.

"I will now state what is my idea of a perfect Silver-spangled cock, beginning with colour, as that is of the most importance. The ground-colour must be a clear silvery white, perfectly free from yellow tinge. I speak, of course, concerning birds in full and perfect plumage; as very many birds at the end of the season, if they have been exposed to the weather and sun, will become yellow. The spangles should be a rich satiny green-black, and their form (on all those parts of the body of the cock which show the full size, as, for instance, the breast and tail) as round as possible. The disposition of the markings is as follows: breast well and boldly spangled from the throat down to the thighs and fluff, black fluff being an especial abomination. The larger the spangles are the better, provided only that a sufficiency of white is shown, that is, if, looking at the breast, both black and white appear distinctly. In some the spangles are so large that they overlap, and give the breast the appearance of being black; this, of course, is a fault; the spots should be as large and round as possible, but so as to show the white between. The neck-hackle is white, but if spotted at the bottom all the better. Back and saddle-hackles should be well spotted with black. The bars on the wing, formed by the large spangles on the end of the primary and secondary wing-coverts, are two in number, and should be bold and regular; these bars are one of the most cardinal points. Above the bars, or the wing-bow as it is called, should also be well spotted; it can scarcely be called spangling, as the feathers in this part of the cock are different to the hen's, being long and narrow. This remark also applies to the back and saddle; I therefore use the word spot to express the marking, instead of spangle (see Fig. 83). The 'stepping' on the wing secondaries should also be well defined—that is, each feather should have a very bold crescentic spangle at the end, which gives an appearance of black steps. The tail to be clear white, with a large bold spangle at the tips of the feathers; though a little colour in the hanging or side-feathers is not objectionable, provided the sickles and secondaries are clear. The comb should be even, firmly set on the head, long, and moderately broad, full of 'work' or points, free from hollow in the centre, and ending in a long pike slightly pointing upwards. The beak should be horn-colour, ear-lobes a clear white, smooth, and as nearly round as possible; face red, quite free from white; and eyes, in this variety, a dark hazel. The legs are slaty blue. As regards shape and carriage, the neck should be nicely arched, with very full hackle falling well on to the shoulders; the breast full, broad, and prominent; back a moderate length, broad and level across, not round or up at one side; tail full, the sickles long, broad, and well arched, and the side or furnishing feathers nicely arched also, the whole to be gracefully carried, not squirrel-fashion, but very slightly drooping behind the perpendicular line, and to be evenly set on, not carried on either side. The whole carriage to be graceful, jaunty, and cheerful. Size, say about five pounds, but this is not of great moment, provided he is not very small.

"The hen should be boldly and evenly spangled throughout, the spangles being round and

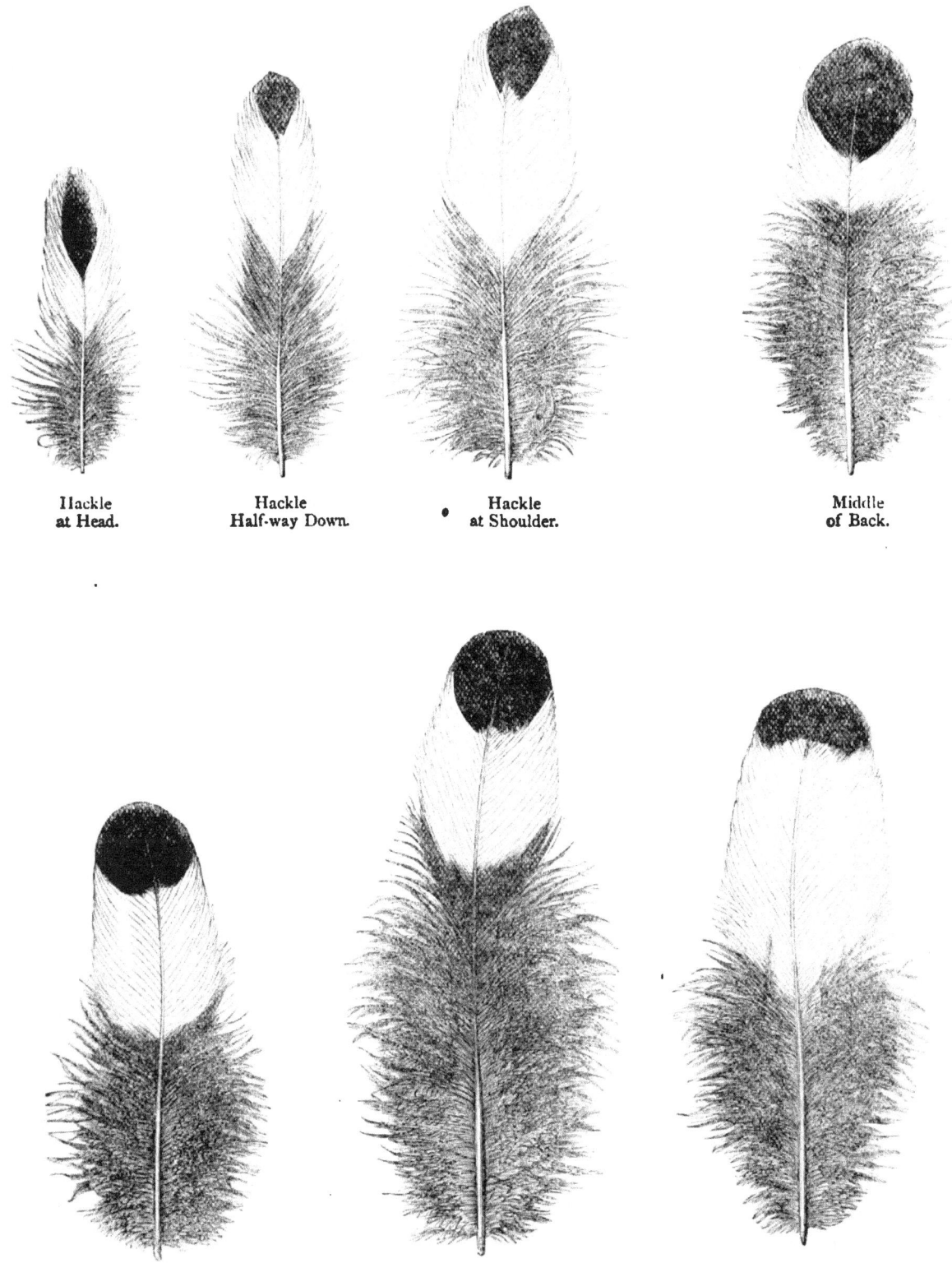

Fig. 84.—Feathers of Silver-spangled Hen.

large, but not of such a size as to overlap each other and give the bird a patchy appearance; on the contrary, the white should show between each feather, though a bird *perfect* in this respect is scarcely ever seen. The colour of the black spangles is a very great point; it must be a very rich satiny green-black, the spangles having almost a raised appearance; in fact, I have taken hold of some hens on which I have almost fancied that I could actually *feel* the spangles. The ground-colour is a perfectly clear silvery white, and the marking as follows: The neck to be well spotted from the head downwards; back well and evenly spangled (I like a hen with a broad back, as there is then more room for the spangling, which appears to better advantage); wings well spangled with bold spangles above the bars, which bars are double as in the cock, and must be bold and well defined; breast spangled from throat down to fluff; tail clear white—no pepperiness in it— and with a good bold spangle at end of each feather. The comb should be smart, full of 'work' or points, with a long spike behind, evenly and firmly set on the head; the ear-lobes white, beak horn-colour, legs slaty blue, and eyes hazel as in the cock. For size say about four pounds, though this is not of importance if they look a pretty fair average.

"For use I consider this variety of fowls to stand in the very first rank. No breed *can* excel them in laying properties, and many birds lay eggs of a pretty fair average size, while the food they consume bears favourable comparison with that required by other breeds. They need a good grass-range to fully bring out their best qualities, though I have known this variety do pretty well in a confined yard; still it seems really a pity to pen them up, and I do not advise any one to keep them who has not a wide run. For beauty no fowl can excel them, in my opinion. They are very prolific, the eggs hardly ever failing to hatch, and the chickens are very lively from the first, and easy to rear. They reach maturity pretty early, and if well housed will lay at from five to six months old."

When the Yorkshire Pheasant and Lancashire Mooney were bred together to produce exhibition cocks, the practice was to mate Mooney cocks with Pheasant hens, and some breeders still pursue this plan, breeding the Mooneys pure for their pullets. We may, however, remark that though old fanciers can distinguish pure Mooneys from crossed birds, it is almost impossible for a beginner to do so, and the pure breeds are now very difficult to obtain, except by the inhabitants of their native counties, who know where to lay their hands upon them.* Should any of our readers, however, be able to obtain a pure Mooney strain, and be desirous of breeding from it for the sake of the pullets, while yet wishful to breed cockerels also for exhibition, and only having one run, they must proceed upon the following plan, which has been followed under Mr. Beldon's advice with success. Get a pure Mooney cock (if whitish in the ear-lobes all the better), and of course pure Mooney hens also for the pullets. Then add a couple of hens for cock-breeding, with very smart combs and good ear-lobes, and very clearly marked, especially about the tail, but somewhat lighter in the spangling. From such hens and the Mooney cock it is

---

* In a note received since the above was written, Mr. Beldon added, "Nearly all the so-called pure Mooneys now have white ear-lobes, showing they have been modified by breeding. Hen-feathering is no sign of absolutely pure blood, being very easily produced. In Yorkshire also, though I live there, I should have some difficulty now in finding a pure-bred Pheasant, our Mooney hens having a dash of the Pheasant, and our so-called Pheasants a *lot* of the Mooney. At the last Birmingham Show (1872) there was not one *absolutely* pure Mooney hen, though there were some splendid birds. Some ten or twelve years ago I came into possession of a lot of Silver Mooney hens, the really pure old stamp, picked up from all parts of Lancashire by old Jack Andrews [referred to on next page]. These hens were much larger than those we have at present, and were certainly coarse; but for spangling—it was perfection! Still, I think among the amalgamated strains we have as good, and that moult as true. I inquired of old Jack, a very short time ago, if he thought any of this 'old sort' could be found still; he said he had looked the whole county through, but could find none."

GOLD-SPANGLED HAMBURGHS. SILVER-SPANGLED HAMBURGHS.

very likely good show cockerels may be produced; but the mode of breeding recommended by Mr. Beldon is preferable in many ways, and will very speedily be the only method practicable.

From feathers kindly supplied by the same authority we add representations which will make clear the difference in the strains, and especially in the marking of the hen-feathered Mooney cocks. None of these feathers, or any others in this work, are in the least idealised, but faithfully drawn just as they appear upon the birds.* The clear spots on the hackles of the hen-feathered cocks will be noticed, and also the substitution of round black spangles on the shoulders and back, instead of the pointed character of the plumage in the full-feathered cocks. We also include amongst the hen-feathers a figure of the old pheasant marking, which, on account of its crescentic character, is now carefully avoided by all good breeders.

GOLDEN-SPANGLED HAMBURGHS.—"In the early part of the poultry-showing era," continues Mr. Beldon, "as in the case of the Silver Spangles, there were two varieties also of the Golden-Spangled Hamburgh, the one most common at the poultry-shows being however in this case the Yorkshire breed, known as the Golden Pheasant. These were fine large birds, and the cocks as a rule hen-feathered; in fact, at this time the hen-feathered birds were all the fashion, so that, although there did not lack full-plumaged birds of the same variety, only the hen-feathered were retained. This Yorkshire Pheasant breed generally produced capital layers. The spangling was bold and of a glossy green black, but the ground-colour was of a light dull bay, and generally there was a good deal of what I may call pepperiness in the ground-colour, so that the spangling was often not clear and sharp-looking, especially in the tail-coverts. As a rule this variety had whitish ear-lobes—not of course so white as we have since bred them by judicious crossing, but still whitish, and besides being, as already stated, a good layer, was a pretty hardy breed.

"In Lancashire there was another variety, cultivated chiefly by the weavers and colliers. This was called the Golden Mooney; it was a much smaller bird, but for colour and marking threw the Yorkshire Pheasant entirely into the shade. I shall never forget my feeling of pleasure on first seeing the Golden Mooney hen: she struck me as something wonderful. The ground-colour of the plumage in these fowls is of the very richest bay, the spangling very bold and clear, and of a green satin-looking black; in fact, the plumage was so rich and glossy that the full beauty of it could not be seen except in the sunshine, but when it *was* seen formed a picture never to be forgotten. I am here speaking of the hen. The cocks' plumage was also of the very richest description; but their great drawback was their red ear-lobes and black breasts—in fact, they had no ear-lobes at all to speak of, but merely a bit of red skin like a Game cock. These cocks were never shown, but merely kept for breeding purposes. Shows were held in many of the village public-houses in Lancashire, the competitors being mostly colliers and weavers of the district, to whom is entirely due the credit of bringing the celebrated Mooney marking to such perfection. At these shows hens only were shown, of both Golden and Silver Mooneys and Black Pheasants, but far the most usually it would be one of the Mooney breeds. The birds were judged by a Scale of Points brought out by members of these village clubs, and the points were so well understood by all that any disagreement about the judging scarcely ever took place. These village shows are now things of the past, the poultry-shows held in almost every town in Lancashire making them unnecessary; but they did work it is very difficult to estimate now. One of the foremost men at these village clubs was old Jack Andrews, or, as they call him in Lancashire, 'The Ould Poo't,' meaning 'The Old

---

* We make this remark because letters have reached us complaining that some of the figures of feathers we have given represent an unattainable standard. Such is not the case; in no instance have we permitted the least exaggeration.

pullet.' What precise meaning attaches to this soubriquet I never inquired, though it is evidently connected somehow with the old man's triumphs at these shows where only hens and pullets were shown; but he always takes it in good part. The old fellow is a rare breeder still; but the Hamburghs—Gold and Silver—are the only varieties he cares about, and I much doubt if he could tell even the names of many other fowls. Another fine old fancier and breeder is old Nathan Marlor; and I must say that breeders and fanciers of both the Spangled varieties are much indebted to these two men, who have been greatly instrumental in bringing the Mooney to such a state of excellence.

"Both the breeds thus described being in existence at the early time we have been speaking of, the Lancashire Mooney hens were first shown with the Yorkshire hen-feathered cocks; but when the judges began to set their faces against the hen-feathered birds (and their reign was very short), the Yorkshire Pheasant's career, as an exhibition bird, was over. After that the cocks shown with the Mooney hens were of the full-feathered Yorkshire breed; but neither variety then possessed the points of excellence required by the judges. The Yorkshire Pheasant was too dull in ground-colour and not distinct enough in the Spangling; while the Mooneys, especially the cocks, had black breasts and red ear-lobes, and if anything (especially in a room) were almost *too* deep and rich in the ground-colour. The necessary change began first with the cocks. As the Yorkshire birds were found to be too dull in colour, and it was out of the question exhibiting Mooney cocks, the Yorkshire cock was put to the Mooney hens, and thus by careful and judicious crossing a bird was produced having somewhat of the richness in plumage of the Mooney, at the same time retaining the spangled breast and whitish ear-lobes of the Pheasant. These were the cocks for some time shown with Mooney hens. After a time, however, as the competition became keener, and richness of plumage became one of the chief points requisite to success, a little more of the Mooney blood was introduced; and at the present day we possess cocks which leave little to be desired either in that point or in ear-lobes, which have been by careful breeding brought to a perfection neither breed originally had. The hens also were operated upon, the red ears of the pure Mooney being found an eyesore. To remedy this a dash of the Yorkshire Pheasant blood was introduced, which also had the good effect of giving a very slightly lighter tint to the ground-colour; and now, by this crossing and judicious selection, we possess a breed of hens also that combine all the richness of the Mooney with a slightly lighter ground-colour and the desired white ear-lobe. Both sexes thus containing some mixture of blood, are gradually approximating; and though I am bound to confess that the most successful breeders still use two pens to breed from, the distinction will gradually lessen, and we are rapidly approaching in this variety also to a strain which will breed both sexes without more difference than all varieties usually exhibit in point of excellence.

"As in the preceding variety, breeding from the pure Yorkshire Pheasant is now almost discarded. For breeding pullets I advise the beginner to get the very best hens from an exhibition point of view that he can lay his hands upon; there is not much need to inquire about the strain, as in a hen this will speak for itself. Then let him get a cock from some good breeder out of a well-known pullet strain, and if possible of the very same strain as the hens or pullets he is breeding from; for I always find that birds bred akin produce by far the most perfect specimens. Then if you find these birds produce first-class chickens, keep them together and breed from them as long as ever you can, and do not on any account attempt to improve them by a cross, or you may improve them the wrong way. If the produce is not satisfactory, try again; but this simple method of selection will rarely fail, and is about the only one which can be given at the present day.

"For cockerels pursue a similar plan. Get the very best exhibition cock you can procure, and put him to a hen or hens obtained from some good breeder; but in choosing them you must select birds with the necessary points, viz., with smart, even combs, and pure white well-shaped ear-lobes. Here, as before, if the produce is good stick to it so long as the pen will breed; but if you do not succeed at first you must change the birds until you get what you require. Any one without such patience and perseverance will never make a fancier; and while we use our very best judgment of course, it is greatly by this experimental method the best of us make up our Hamburgh pens, until we have got a strain of our own, when of course we know its qualities, and can make up our pens with something like certainty of success.

"I will now describe what Golden-spangled Hamburghs ought to be, beginning with the cock. It will be understood that in choosing stock as above described for breeding either sex, the most important sex in each case is to be chosen as nearly approaching the following description as possible.

"The cock's hackle should be a rich golden bay, each feather striped down the centre with rich deep black; the back a deeper bay, approaching maroon, and each feather having a green-black spot, these spots getting more elongated as they get down the back, till on the saddle the hackles become striped down the centre. The red or maroon should be very rich in colour. The breast is golden bay, each feather well spangled with a rich black moon; this spangling to be very uniform, and not to overlap, but to show both black and bay; the breast to be thus spangled from the throat to the thighs, the spangling becoming bolder as it goes downwards. A laced breast is objectionable (although many of such birds win simply for want of better), but a black breast is now out of the question. The tail should be black, very full and long, and as rich in colour or green gloss as possible. The bars on the wing should be double, bold, and regular, as in the Silver-spangled breed, thin and imperfect barring being a great fault; the stepping on the wing, caused by the black crescentic spots on the ends of the secondaries, should also be good. The outer webs of both secondaries and primaries to be a deep golden bay, but the inner webs ought always to be black. The wing-bow, or part above the bars, should be rich maroon, with each feather spotted with black if you can get them so; this, however, is very seldom the case, still the wing-bow should at least be very rich, and the ends of the feathers darker, approaching to black. The comb resembles that of the preceding breed, and should be smart, full of 'work' or points, with a long spike behind slightly pointing upwards, set evenly and firmly on the head, and quite level on the top; any hollow in the middle being a great fault, as the comb is one of the first points which catches a judge's eye. The ear-lobes should be not only a clear white, but of a nice shape—as round or circular as possible, and not long or pendent, but nicely put on the face. The face is red, beak a dark horn-colour, eyes bright red, legs a dark slaty blue. The carriage is easy and graceful. Size is not of much consequence if not very small.

"The ground-colour of the hen is a rich golden bay. The neck-hackle is of the same colour, each feather striped down the centre with deep green-black. The breast from the throat to the thighs should have each feather spangled with a bold, rich, round, black moon. The tail is black, the coverts being spangled. The back should be broad, and richly spangled, and the shoulders or bows of the wings above the bars should especially be well spangled. The bars themselves are, however, the chief point, and should show plainly as two bold and regular lines of spangling, the want of which is a most serious fault, and mars altogether any other beauty in marking. The comb, face, eyes, &c., resemble the cock's; and, as in his case, size is not material provided the bird be not too small.

"Golden-spangled Hamburghs are only moderate layers in comparison to the other varieties, the pure Golden Mooney on which they have been founded being very indifferent in this respect.

"Spangled Hamburgh chickens vary in colour, the Silvers particularly so, according to the different strains, or even sometimes in the same strain. Some hatch out a light grey, others a dark smudgy grey, and striped black and grey down the back and sides; while of the Golden some are dark brown striped with black, and others very light, almost yellow. In the first feathering they also vary much in colour, some being very light—that is, the white predominating—and others dark or almost black. However this may be, in the first feather it is a blotchy black and white, without any true spangling. Many have their wing-feathers pencilled, others not, this being not an invariable rule. There is, however, never any true spangling in the first feathers, and as a rule the darkest birds in their chicken dress prove the best; but there are exceptions to this too, so that if the strain can be depended on it is best to wait for the second or adult plumage. Then it is that the true character of the bird is developed, the difference being wonderful. Now all becomes distinct and well-defined; and as the new feathers grow it can soon be seen whether the chicken possesses the requisites of an exhibition bird or otherwise."

As a rule, it will be found that the cocks referred to by Mr. Beldon, to be selected from a good pullet-breeding strain for breeding pullets, are somewhat darker in ground-colour and coarser about the head than good exhibition birds. If possible, own brothers to winning pullets should be selected; and as soon as a good strain has been formed the amateur will of course select cockerels from his own pullet-breeding strain, and *vice versa*. In no variety, therefore, is it so necessary as in either of the Spangled Hamburghs to form as soon as possible a strain of one's own.

Originals of the Scales of Points for judging Golden Mooney hens, referred to as in use at the old village shows, are now very difficult to obtain, though they were printed for reference. They have, however, been fortunately preserved in the original "Poultry Book," now rather scarce, published in 1853 by Messrs. Wingfield and Johnson, from which (by permission) we copy them, as most interesting to all fanciers even of the present day. The same table applied to the Silver Mooneys, substituting a white for the red ground.

POINTS.	MARKS OF FEATHERS, ETC., CONSIDERED BEST.
1st.—COMB	Best double; best square; the most erect and best piked behind.
2nd.—EARS	The largest and most white. [White, and a medium size. N. M.]
3rd.—NECK	The best streaked with green-black in the middle of the feathers; and best fringed with gold at the edges.
4th.—BREAST	The largest moons; best and brightest green-black, most free from being tipped with white or red [omit "or red." N. M.] at the end of the moon, and the clearest and best red from the moon to the bottom-colour.
5th.—BACK	The largest moons; best and brightest green-black, least tipped with white or red [omit "or red." N. M.] at the edges of the moon, and the best and clearest red from the moon to the bottom-colour.
6th.—RUMP	The largest moons; best and brightest green-black, least tipped with white or red [omit "or red." N. M.] at the edges of the moon, and the best and clearest red from the moon to the bottom-colour.
7th.—WING	This is divided into four parts:—1st, *Bow.* Best and brightest green-black, and best and clearest red.—2nd, *Bars.* To have two distinct bars, composed of the largest, clearest, brightest, and best green-black moons, and the clearest and best red from the moon to the bottom-colour.—3rd, *Flight.* The clearest and best red [quill of the feather to be same colour as the web. N. M.].—4th, *The Lacing, or top of the wing, above the flight* (now called "stepping" of the wing. Largest, clearest, brightest, and best green-black spots on the end of the feathers, and the best and clearest red from the spot to the bottom-colour.
8th.—TAIL	The brightest, darkest, and best green-black. To be full-feathered.
9th.—LEGS	Best and clearest blue.
10th.—GENERAL APPEARANCE	The best-feathered hen.

In both varieties hens only were shown, the cocks being so inferior in appearance as to be only valued for breeding. The spangling was really marvellous in its gloss, as we can testify from feathers kindly sent us by Mr. Nathan Marlor, the breeder referred to by Mr. Beldon, and who also states that the corrections within brackets bearing his initials should be made in the scale as given by the "Poultry Book" to make it accord with the views of the best old village breeders.

The tendency to which we adverted at the commencement of this chapter towards a white or red tip beyond the moon, thus showing a distinct approach to pencilling in character, will not fail to be noted as clearly pointed out by these old rules. Golden Mooney hens as they become aged not unfrequently show this slight white tip on the ends of their feathers beyond the moons. Occasionally the tip is bay; but in neither case is such necessarily an indication of impure blood. A laced character, on the contrary, especially on the wing-bars, shows not only too much Pheasant blood, but that it has not been properly amalgamated—in fact, that birds showing it, if adult (chickens of very good quality sometimes do show a little lacing, but afterwards moult out correctly) are in reality badly-bred birds, and have not been produced with the care which should have been observed, as described by Mr. Beldon. The amalgamation is still going on, having made perceptible progress during late years; and so far as can be foreseen, in a very few more we may expect to see a thoroughly established and improved strain, which shall breed both cockerels and pullets alike with all the points desired, though no doubt to the very end certain families will produce better of one sex than the other, as is generally the case with every other breed. Even this, however, is in the hands of fanciers; just as we have made already a large approximation towards amalgamating the black breast now demanded for Dark Brahma cocks with the necessary pullet-breeding qualities in which that kind of cock was formerly deficient, and which therefore were and are often still sought in speckled-breasted birds.

With the exception of the tail-feathers, which in Golden-spangled birds are black, and the hackles, which are striped, the plumage is so similar to the Silvers that representations of the feathers need not be repeated.

The Yorkshire Golden Pheasant fowl* is now we believe nearly extinct, and even the Mooney hens are not nearly so often shown pure-bred as formerly, the somewhat lighter colour of what we may call the new or amalgamated strain being preferred. The showing of both has given rise to some controversy at times between the advocates of what have been called the red and the golden ground-colour. We think the truth lies entirely on neither side. Under cover, or the conditions of the show-pen, there can be no doubt of the superior effect of the lighter tint, which was deliberately sought by the best breeders for this very reason; but out of doors, especially in the sun, these birds are not in our opinion to be compared with the true-bred Mooney in point of feather, though the smarter heads and white ear-lobes give a great superiority in those respects.

REDCAPS.—Nearly all the older works on poultry describe among the Hamburghs a breed called the Redcap, for which, years ago, there was a special class at the Sheffield shows. They appeared to be a kind of mongrel Golden-spangled, larger in size, and with *immensely* large rose-combs, often hanging over at the sides. They were reputed to be hardy fowls and good layers. The late Mr. Hewitt entertained the very highest opinion of them as "general utility" fowls, as the following interesting notes, furnished by him to the first edition of this work, will show:—"Of the

* It will of course be understood that the Yorkshire Pheasant breeds so frequently referred to have no connection with the pheasant itself, but that the term is merely a local *name*. Hybrids between the pheasant and fowl are common enough, but, like other hybrids, are almost always sterile, and quite incapable of founding a breed. The name was evidently given from the similarity of marking of the breeds in question to those of the pheasants.

Redcaps," he says, "I can without hesitation speak most favourably, both as regards the production of eggs and also their value as a table fowl. I never kept them myself, but have been intimate with several parties who esteemed them most highly, long before the time that poultry exhibitions were instituted in the Midland Counties. To a poultry amateur, whose eye has been previously tutored to the most important traits of character in Hamburghs generally, the Redcap at first sight presents nothing less than a mass of general disqualifications, as such parties very unjustly form their opinions by comparison with the code of rules by which the value of other varieties of Hamburghs are estimated. Although the very profuse rose-comb, lounging in a very ugly manner, the partially pendulous and very red ear-lobe, the just barely crescented feather, in lieu of a spangle, and the want of sprightly motion, so characteristic of all the Hamburghs, are far from ornamental, added to which the ground-colour is anything but as sound as could be desired by the party whose search is exclusively for beauty of exterior, the compensations for these shortcomings are profuse; for they are really a weighty and thick-bodied fowl (cocks reaching seven pounds and a half), of good flavour on the dish, and if the eggs are weighed, as well as counted, I believe them to be the most abundant egg-producers of all our domestic poultry. In reference to their eggs, I will mention a fact to which my attention was first directed by one of the oldest and most practical Birmingham confectioners. If after being broken the same weight of eggs is used from Redcaps and Spanish fowls, the consistency in custards and so forth obtained from the first-named breed proves nearly one-third greater than from those of the Spanish. To such parties as use considerable quantities of eggs for confectionery purposes, this peculiarity of the Redcap's makes them much sought after; and, I may add, each individual egg, when the fowls are well attended, is as fine and noble-looking a specimen as could be desired."

The publication of this high commendation by the late veteran judge perceptibly stimulated the demand for the fowl. It gradually became more inquired after, and now and then a class appeared at a show, which never failed to awaken considerable interest amongst the visitors. Occasional articles appeared in the poultry papers, and it was no longer difficult to procure stock. Mr. A. E. Wragg, of Edensor, Bakewell, was especially prominent in drawing attention to the breed by articles and letters contributed to various poultry publications. To the efforts of this gentleman its growth in popularity has been greatly due; and we are glad to be able to give the following notes from his pen :—

"Of all breeds of poultry perhaps not one has been more misrepresented or has received less encouragement than the Redcap. Only recently have its merits obtained general recognition, but it now promises to become one of our most popular varieties. Many have written disparagingly of Redcaps who have never seen a pair of well-bred birds, describing them as simply mongrels with a cross of the Golden-spangled Hamburgh in them. A writer in one of our weekly periodicals, some years ago, declared that they might easily be bred by simply crossing almost any kind of common barn-door hens with a Golden-spangled cock, and no doubt many people believed him.

"Although one of the oldest of British breeds, until lately very little was known of it except in Yorkshire and Derbyshire. It is generally supposed to have originated in Yorkshire, many believing it to have been produced from the Old English Game and Golden-spangled Hamburgh. There is little doubt that Game blood does enter into its composition, for there is a large amount of Game spirit in the breed, and, in fact, a Redcap cock when dubbed might almost be taken for a Game cock. The breed has gone under many different names, such as Pheasant Fowls, Moss Pheasants, Crammers, Copper Fowls, Yorkshire and Derbyshire Redcaps. My present strain has been produced from the two latter varieties.

I am informed that it was in Nottinghamshire that the name of Crammers came into use. With regard to its origin, my own opinion is that the Redcap is the original of the Golden-spangled Hamburgh, and this opinion is shared by many of the modern poultry writers.

"The Redcap cock is a fine-bodied bird of noble and commanding appearance, the 'crowned' king, as it were, of all poultry. Nothing could possibly be more ornamental than his large symmetrically-shaped comb, full of a great number of fine, long spikes, with leader behind. It should be well carried—firm, straight, and standing well up from the eyes. For

REDCAPS.

years Redcaps were bred with very ugly combs; and to this fact may be attributed much of the unpopularity of the breed. The improvement in the Redcap comb during the last ten years is something wonderful; an ugly comb being now very rarely met with. With regard to its size, it should be as large as can be comfortably carried by the bird. The comb of an exhibition bird which I measured was $5\frac{1}{4}$ inches in length, and $3\frac{3}{4}$ inches in breadth, but I have had birds with even larger combs than this. The weight of this cock was $7\frac{1}{4}$ lbs. I exported a cock two years old to Australia in March, 1889, which weighed just 9 lbs.

"The principal faults found in combs are—size; want of spikes; being badly spiked behind; having a hollow in the middle; unsymmetrical shape; being too heavy in front, falling to one side, or hanging over too near the eyes. A cock which is very faulty in any of these points will stand scarcely any chance of winning in good competition, if well judged. The comb should be very red, and this redness cannot be obtained unless the bird be in first-class condition. In no other breed is condition of more importance than in this. We often see birds of other varieties winning when in bad condition; but I have never yet seen Redcaps, shrunken in comb and rough in feather, in the prize list. On the other hand, the very best birds in poor condition are often beaten by the most worthless specimens, red in comb and glossy in plumage.

"Some would-be fanciers advocate that smaller combs should be bred, but I cannot think that any true Redcap fancier would care to see them with Hamburgh combs. The large comb is, and always has been, for the last hundred years, the distinguishing feature of the breed, and must on no account be done away with. If they are to be bred with small combs, then let another name be found, for they will no longer be Redcaps. The large comb is not the creation of to-day, as many appear to think; they have always been bred with such combs. The large comb does not look out of proportion to the size of the body, and there is not one bird in a thousand that cannot eat, fly to its perch, and carry its comb as comfortably as a Minorca.

"The ear-lobes are red, neck and saddle rich red, striped with black. Many cocks are too black in neck hackle; others are spoiled by being too yellow. The back of the cock should be red, spangled with large half-moon black spangles, breast and tail black, legs slate-coloured, strong, and of good length.

"In general appearance the Redcap cock is much like the Hamburgh, being smart and lively; the breast full and round; tail ample and well furnished; and shape symmetrical and pleasing to the eye.

"The Redcap hen is a large, round, comfortable-looking bird, weighing about 6 lbs. on the average. Good exhibition specimens often weigh considerably more than this. Two hens I have just weighed turn the scale at $6\frac{1}{2}$ lbs. and 6 lbs. 10 oz. respectively. The ground-colour of breast, back, and wings is a deep, rich, reddish brown; each feather being tipped with a half-moon spangle of bluish-black, and these spangles should appear as regular and uniform as possible. Redcap hens are, many of them, too light in ground-colour, and, instead of being spangled, are laced with a thin edging of black. These birds have been produced by mating with cocks having yellow neck hackles. The comb of a good exhibition hen should be like that of the cock in everything but size. The comb of one of my best hens is $2\frac{3}{4}$ inches by 2 inches when she is in full condition.

"There is no breed that can surpass the Redcap for laying qualities. As a rule, they will average about 180 eggs in the year, *without* exercising any particular care in the selection of the best layers, and many hens in their second year will lay more than 200 eggs. I have now a breeding pen of four hens, one five years old, two four years, and another two years old. Some of them are exhibition birds. The first began laying this year (1889) on January 27, the remainder in February, and up to the present time (June 26) they have laid 413 eggs, showing an average for the five months of 103. Hens of two, three, and even fours years of age will generally lay quite as well as pullets, and no hen should be killed until she is three years old, unless she has been proved to be a bad layer.

"In my opinion the Redcap egg is the richest laid by any variety. The average weight of each egg is about 2 oz., and the colour white or slightly tinted. From a box containing about a score, I selected six of the largest, and found the weight to be exactly 13 oz., showing an average of 2 oz. 5 dr. for each egg. I have kept Redcaps for the last ten years, and, far from having noticed any deterioration in their laying powers, am quite certain that my birds now are far better layers than the stock I possessed at the commencement.

"The chickens are hardy, easy to rear, and feather fast. With a good grass run, and plenty of good sound grain twice a day, they come on very fast, and scarcely ever ail anything. My birds are reared on very cold exposed runs; and some generally roost in a plantation all through the winter, just like pheasants. As a rule, pullets are not very precocious, and do not usually commence to lay before they are seven or eight months old. Those hatched in March and early in April, if well fed, will lay in October and November, but if hatched from May to July, will generally begin in February.

"The general improvement in the Redcap has been frequently noticed in reports of shows held during the past two or three years. A show was lately reported of as follows :—' The winning cock is especially good in comb and shape, and the first prize hen is *as good in spangling as a Hamburgh*; the classes were both very fine, and the breed is certainly improving.'

"It is a pity the breed is not more encouraged by show committees, and I could never make out why it is not, as a class is almost invariably well filled, and constitutes one of the principal attractions of the show."

Information from other sources has brought to light other curious synonyms for this breed besides those mentioned above. They have been known as "Manchesters," and also as "Poland Pheasants." We have been informed that many who kept them in old days were so jealous of their egg-laying capacities becoming known, that they pricked the eggs sold, lest purchasers should obtain any of the stock so highly valued. We have traced this practice to several quarters, and the remarkable jealousy thus displayed is worth recording, as it easily accounts for the otherwise strange fact, that over and over again Redcaps have disappeared from districts where they were once kept. But wherever the fowl was known it was highly valued. Quite lately one farmer has reported a cross between a Redcap and a Partridge Cochin as a fine *table* fowl—"the best he had seen in his life;" and the same farmer averred that one hen he had, now in her eighth year, has laid more eggs during the last twelve months than at any previous period. This testimony amply confirms that borne by Mr. Wragg to the capacity of the breed for laying on *late in life*, in which respect we believe no other breed is known to equal it. No doubt the somewhat late commencement has a great deal to do with this continuance of laying powers.

Perhaps the points chiefly to be questioned are concerning that very "improvement" mentioned in the above sentences from a poultry journal. Undoubtedly the Redcap *can* be bred up to a beautiful spangling, and a neat straight comb, and the process has been carried already to a certain point, while judges are evidently awarding prizes very much for such points. But the question arises, What will the Redcap be when the process is finished? Will it be the hardy fowl and magnificent layer which was known before? We very gravely doubt it. The breed does not seem to have suffered yet, but the process has gone on only a very few years: it will be different when in-breeding and selection for exhibition points have done their work for another ten years, and the fowl has become to all intents and purposes a large Golden-spangled Hamburgh. A large comb, in particular, seems to have some connection with laying properties; and we may be allowed to express an earnest wish that a most valuable fowl may not be spoilt by insisting in the show-pen upon qualities which do not belong to it.

No Club being yet formed, no standard has been adopted for the Redcap fowl; but our own strong opinion is that symmetry, size of body, and size of comb, ought to be the chief points, if the useful qualities of the breed are to be kept up.

SILVER PENCILLED HAMBURGHS.—"This breed and the following," observes the same authority already quoted, "are somewhat smaller and lighter in make than the Spangled varieties. In Yorkshire it is often known by the name of Chittiprat, and in Lancashire by that of Bolton Greys, but these and all other names are gradually giving way to that of Silver-pencilled Hamburghs. I will first describe the proper markings of the breed, as formerly sought.

"In the cock, the head, hackle, back, saddle-hackle, breast, and thighs should be a clear silvery white, the yellow tinge which is so often seen being a very grave fault. The tail proper is black, the sickles and side or furnishing feathers being a rich green black edged with a fine white fringe, the finer and more sharply defined the better. Many birds have a sort of marbled tail, which is very objectionable; others have the sickle-feathers splashed with white, which is also a grave fault, the only white which should be in the tail being the fine white edging merely, which is clearly shown, with the other important markings of the cock, in the plate (Fig. 85)—the sickle is drawn more bent than usual, in order to get it within the plate. The wing appears almost white when closed; but the inner webs of the wing-coverts should be very darkly pencilled when examined (for which again see Fig. 85), and a fine black edging should be observable on the wing-coverts, caused by the ends of the *outer* webs being also slightly tipped with black, and giving the appearance of a slight and indistinct bar on the wing. If this point is not observable it is a great fault, and such birds should never be bred from; on the other hand, this barring should not be too distinct or heavy, as such gives the bird an appearance (very often correct) of being crossed with a coarsely-pencilled strain. The colour of the secondary quills is also important. On the outer webs they are white, except a narrow strip of black next the quill, which is of course only discernible when the wing is opened out, the white only being seen when closed; the inner web is black, all but a narrow white or greyish edging. In some birds the colour is a sort of black and white mixture or marbly appearance, and not nearly deep enough, especially on the inner web. Such birds have been bred from indifferently or lightly-pencilled hens, and are of no use at all to breed from, at least if you expect to breed pullets as well as cockerels. On the fluff of the thighs are some black spots or pencillings; but with this exception and those already stated, the entire plumage should be white. The comb, ear-lobes, legs, &c., should be much the same as described for the Spangled, but somewhat smarter in appearance. The eyes to be a bright red.

"The neck-hackle of the hen also should be pure white, free from spots or pencilling, though hackles somewhat marked are very common. The rest of the body should have each feather distinctly marked or pencilled across with bars of black, as clear and distinct as possible on the white ground, and in particular as straight across the feather as possible. The 'finer' this pencilling, that is, the more numerous the bars across, the better. This pencilling should extend from the throat to the very end of the tail. A tail well-pencilled is a very great point, as there is a special tendency in the long feathers to lose the straightness across of the marking; but tails pencilled squarely across to the very tip can be and are bred, though never common. A very usual fault is a light breast, or if not light only covered with large horse-shoe markings, both being grave faults. The birds best marked on the breast are often a little inclined to be spotted on the hackle, and this latter fault is certainly to be much preferred to a bad breast. The best marking on the breast is never however quite equal to that on other parts of the body. The proper marking of the pullets is shown in Fig. 86, which is drawn carefully and without any exaggeration, from actual feathers.

"The pencilling or marking, as said before, should be as fine and as straight across the feather as possible; and especially, the rows of pencilling on one feather falling on to the rows on

Sickle Feather.   Wing-covert or Bar.   Secondary Quill.

Fig. 85.—FEATHERS OF SILVER-PENCILLED COCK

the next, so as to give the bird a ruled or lined appearance, which has a very pretty effect. Of late years the pencilling of Silver Pencils has failed in being much too coarse.

"The pencilling is generally much the best the first year, or as pullets. Afterwards, as a rule it becomes somewhat mossy, cloudy, or indistinct, and often coarse. Some birds, however, will moult out well the second season, and such should be specially valued, and by all means retained for breeding.

"In breeding this variety, when cocks were selected as described at page 382 and figured on page 383 the same pen would produce both sexes of good quality. Of course there were always

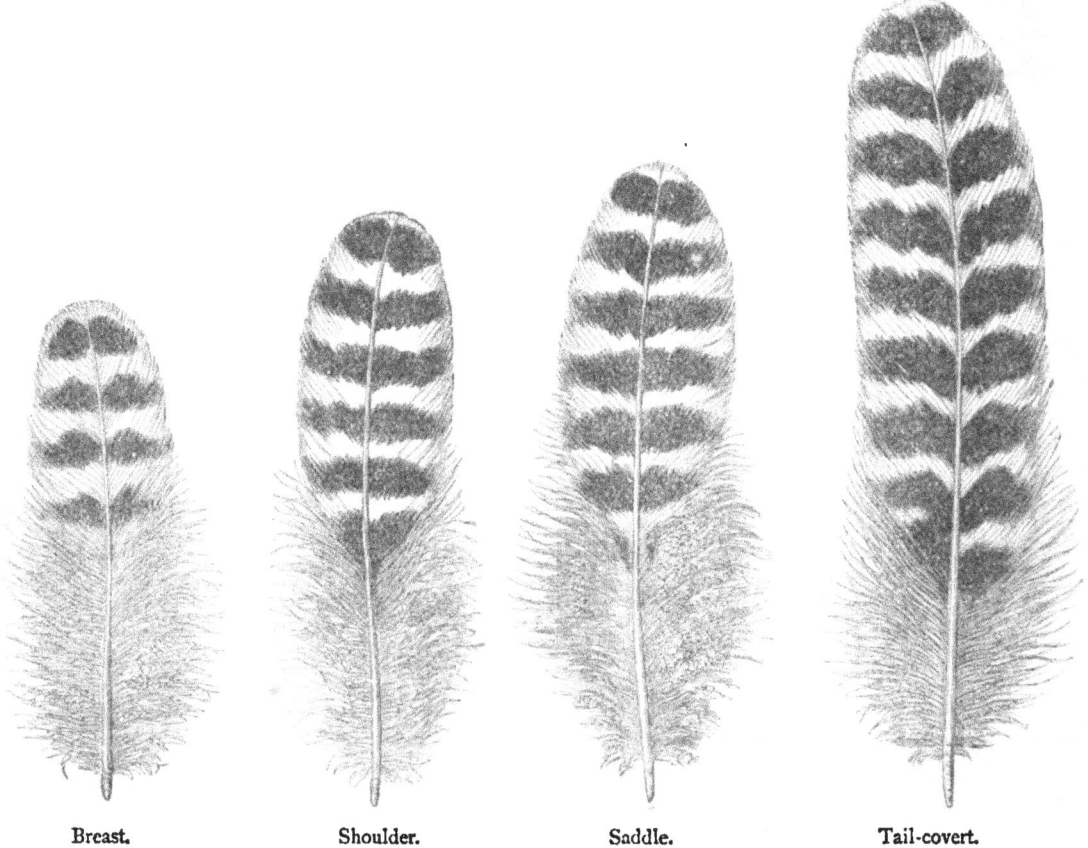

Breast.   Shoulder.   Saddle.   Tail-covert.

Fig. 86.—Feathers of Silver-pencilled Hamburgh Pullet.

some families that produced better of one sex than of the other, as there are in all varieties occasionally, and hence some breeders always used two yards; but others preferred to breed from one set only, and there was no difficulty in it while the judges gave the preference in the show-pen to those cockerels with the points which were likely to produce good pullets. These points were—for the tail to be black, the sickles jet-black all but the edging, and the wing properly barred, with the secondaries also properly edged. But of late years all this is changed, owing to a rage for pure white bodies in the cocks, till the Silver-pencilled Hamburgh now practically consists of two varieties, bred as follows:—

"For breeding cockerels, hens now have to be used as nearly destitute of marking as possible except on the tail—many, indeed, are nearly white. These hens have to be procured from a reliable source, as with the exception of comb and ear-lobes there is really little to distinguish about them. To purchase merely the whitest birds will not answer, as they may breed very bad tails. All depends upon the strain. Such hens or pullets are mated with exhibition cockerels; and these now have to be as free as possible from any marking whatever all over the body, the only black desired being in the tail.

"For pullet-breeding, on the contrary, the breeding a strain distinct for marking has resulted in the production of *hen-feathered cocks* marked very nearly like the hens, with short, henny tails. This is a perfectly natural result, and has followed the constant selection of cocks with the most pencilling on the wing. One advantage of these birds is that the quality of their pencilling can at once be seen; and hence the mating is a comparatively simple affair."

We cannot say that the pencilling of Silver-pencilled Hamburghs has been in our opinion at all improved by the change in the mode of breeding, but rather the contrary. The numbers lately exhibited also show a most significant decline, and we think it a matter to be deeply deplored that the judges so unadvisedly discouraged birds suitable for pullet-breeding. It is not the first occasion, however, on which the exaggerated demands of judges have worked mischief. We may, perhaps, remark that the production of hen-feathered cocks is a strong corroboration of our views respecting the original unity of the Hamburgh race, and shows conclusively how the hen-feathered Spangled cock was produced: it also explains how the hen-tail arose in the Sebright Bantam.

First-rate Silver-pencilled cocks will sometimes, as they get old, show a chestnut patch upon the wing. We remember one of the very best show birds we ever saw moulting out thus after winning thirty-seven prizes, and the fault used to be common in aged birds; but by rigorously excluding such from breeding, it is now seldom seen. Birds which do show the chestnut patch should never be bred from.

GOLDEN-PENCILLED HAMBURGHS.—"This variety," continues Mr. Beldon, "is in every respect save ground-colour similar to the Silver-pencilled breed. The ground-colour in the hens should be about the colour of gold, as rich and bright as possible; the pencilling being exactly like the preceding variety, as distinct and yet as fine as can be got; that is, as many bars as possible across each feather, provided they are distinct and of a good rich black colour. The neck-hackle, like the Silver birds', should be clear. The cock is of a deeper tint, his colour being somewhat between that of his own hens and of a Red Game cock; it must be neither too red nor too pale, but what might be called very rich in effect. His proper tail-feathers are black, the sickles and hangers rich black edged with brown or bronze, the edging being rather wider than in the Silver-pencilled bird. To have sickles all black is a great fault, and so is a tail bronzed all over, or with scarcely any black in it, but bronzed almost all over the sickles. This last kind of tail is very showy, and used to be rather a favourite with some judges who did not understand Hamburghs; but experience proves that birds possessing it produce very indifferently pencilled pullets, and the judges on that account now throw them out.

"Besides the quality of the black pencilling, which resembles the Silver-pencilled, one of the great points in this variety is the evenness and richness of the ground-colour. Some birds, otherwise good, are very uneven in this point, the ends of the feathers being a lighter gold than the other parts. Such birds, as the season advances, are apt to get still more faded and washed-out in appearance, and, indeed, most birds fade in colour from the effects of the sun; but some hens of a good rich colour retain this much better than others, which, is of course, a great point in their favour. In the cocks the same fault is commonly seen, appearing in the shape of a lighter shade on the ends or tips of the feathers on the breast and under parts of the body; this is to be avoided as far as possible—the more uniform the colour the better.

The breeding of Golden-pencilled Hamburghs is in all respects the same as that first described for Silver-pencilled, allowing for the difference in ground-colour. The same markings on the feathers of the cock described at page 378 are to be sought exactly.

"When hatched, Golden-pencilled chickens are a buff colour, darker than the Silvers, with a few black spots about the head. The Silvers are a pale buff. They show the pencilling in their first feathers, the cockerels being pencilled nearly like the pullets. Some breeders prefer to select a cockerel for breeding in either variety at this stage, and no doubt the character of the pencilling he will breed with good hens can be readily seen in this way. The full beauty of the marking does not however show itself till the second or adult plumage makes its appearance.

"Both varieties of Pencilled Hamburghs are delicate, and should not therefore be hatched before April. In the case of the Silver variety there is another reason for not hatching early, in the fact that if earlier they often moult out like old hens at the time they ought to be laying, and thereby lose that sharp and rich pencilling for which the pullets are almost always superior. It is rather rare, for the same reason, to see females of the pencilled varieties shown beyond the first year, and those good enough must be of unusual excellence.

BLACK HAMBURGHS.—"This variety," Mr. Beldon writes, "had long been known in Lancashire previously to the poultry-showing era, but the fowl we possess at present without doubt has some Spanish blood in its composition. This cross was introduced to obtain the white ear-lobe, which was, and is, so very desirable; but the drawback was the white face introduced with it, and which would show itself for a long time. By careful breeding, however, this has been to a great extent bred out, and the majority of the birds shown at the present day have red faces. Of course, the white face will now and then crop up, even now, but it is a blemish, and one the judges will not tolerate.

"This variety is, perhaps, one of the most useful fowls we have. It lays as frequently as the other Hamburghs, but the egg is much larger, probably on account of its relation to the Spanish. Being a black fowl, it can also be kept in the neighbourhood of towns, provided only there be the good-sized grass-run which is needed to do justice to this as to all the other varieties of Hamburghs. In size it is larger than the others, although different birds vary, and size is not very essential if of fair average and symmetrical in make. In shape they should resemble the Spangled Hamburghs, and not be thin and stilty like the Spanish, a point which always shows a bad descent. The legs are a dark leaden blue the first year, which gradually becomes a slaty blue; the comb, face, ear-lobes, and eyes should resemble the Spangled Hamburghs. The colour is a rich satiny green black, the greener and richer the colour the better; and this green gloss should be uniform, and not showing on the end of the feather only, but throughout the plumage. This beautiful gloss shows itself more in the hens than in the cock, which is contrary to the usual rule. To be seen to advantage they should be viewed in a good light, or when the sun is upon them; you then see that beautiful sheen in which they surpass all other black fowls. Some are purple or raven black, but the colour required is the green black.

"The chickens when hatched are white from the throat downwards to the under part of the body, the rest black. As a rule, they do not become thoroughly black till they get into their second plumage."

Other breeders exceedingly dislike the Spanish cross, though we must say that half the birds exhibited show evident traces of it. The Rev. W. Serjeantson has long been known as a special

SILVER-PENCILLED HAMBURGHS. BLACK HAMBURGHS.

cultivator of this breed in its purity, having never we believe crossed his birds; and as none have been more successful with it in the show-pen, it proves that careful and judicious breeding can do all that is necessary. At our request he has added the following very full notes on his favourite variety, which will fully convey to the reader the means by which his great success has been attained.

"It is only within the last few years that Black Hamburghs have come into fashion. Indeed, ten years ago they were scarcely known out of Lancashire and the West Riding; but for all that they are by no means the 'recent invention' which the editors of some of our poultry periodicals would have us believe. I must confess to ignorance myself as to the origin and early history of the breed, but I am glad to be able to supply my deficiency with the following information which Mr. Teebay has very kindly sent me. He says:—

"'I have known the Black Hamburghs ever since I was quite a boy. I do not exactly recollect whether the Silver Mooneys or the Blacks were the first fowls I possessed, but I believe the Black Pheasants, as they were called here. Both kinds, and also the Golden Mooneys, I have been told by reliable persons were exhibited for prizes, such as copper kettles, &c., more than a hundred years before my time. I have yet somewhere in the attics of my house about a dozen prize copper kettles, and many others have been given away to friends.

"'The Black Pheasants were not formerly so elegant in shape as they are now. Little regard was paid to symmetry, but the most weight was given to the resplendent green shade; the ear-lobe was smaller than it is now, and the face a brilliant red. I have known Silver Mooneys produce chickens almost black, and as the old Blacks were exactly of the same shape as the old true Silver Mooney, I always thought one had sprung from the other. The true Silver Mooney chicken is almost black in one stage of its chicken plumage. The Black Pheasant formerly was much shorter in the leg, in fact, shorter in all its parts than the birds we now see.

"'I do not think that crossing with Spanish has caused this alteration; but that it has arisen in most cases from breeders paying more attention to shape than the old fanciers used to do. There is no doubt some strains have a little Spanish blood in them, but these are generally very coarse in comb, with dark faces, inclined to white below the eye, a drooping ear-lobe, and are not nearly so elegant a bird as the true kind.'

"As Mr. Teebay is admitted to be one of the best authorities upon all the varieties of Hamburghs, his most interesting letter is a conclusive answer to those who say that they are a cross between Golden-spangled Hamburghs and Spanish. No doubt some Black Hamburghs, or rather imitations of them, have been concocted in this way; but the unfortunates who get any of these 'cross-breeds' into their yards will not be long in discovering, in single combs and a general want of fixed characteristics, that they have not got the pure breed.

"It will be seen that Mr. Teebay considers them to be closely connected with the Silver Mooneys, and I think myself that that is very probable. However, as they can show a pedigree of considerably more than a hundred years, it is not of great importance from a practical point of view what was their origin in the first instance; it is enough to know that they are now a firmly established breed, of great beauty and undoubted excellence.

"In shape they should be what their name implies—Hamburghs, and not rose-combed Spanish or Black Dorkings. They often appear at exhibitions in the guise of heavy, square, loose-feathered, Dorking-like birds, with coarse heads and combs; or, on the other hand, long-legged, narrow-bodied, and squirrel-tailed, so that one is not surprised to hear the unitiated spectator remark, 'Why, they are very like Spanish.' They should be *real Hamburghs* in shape, with prominent breasts, legs small-boned, taper, and short, though not so short as to give a dumpy look; they should also stand up well upon their legs, but yet without showing the whole of the thigh, like Game. The head should be small and neat. Judges do not seem to care about the colour of the eyes, but I myself much prefer a full dark eye to a lighter-coloured one. The neck-hackles should be long, flowing well over the shoulders and back. The tail should be long, full, and sound-feathered, carried well up, but not too forward, a squirrel-tail being very objectionable.

"Mr. Tegetmeier has stated that their combs are better formed than those of any other double-combed breed; but that, I think, is an opinion which will not be endorsed by Black Hamburgh breeders, who all find the comb one of their *difficulties*. The combs of pullets especially are apt to '*go over*,' as they approach the period of laying; but still, with care in the selection of stock-birds, this is a difficulty which may with certainty be surmounted.

"The ear-lobe is a very striking feature in a Black Hamburgh. It is allowed to be a little larger than in the other breeds; but it must be pure white, round, smooth, like a piece of a white kid glove, lying close to the head, not long and pendent, not puffy, not wrinkled, nor edged with red.

"The face should be brilliant crimson, almost scarlet, or deep carmine; a dark gipsy-face being a great blemish, and a white face altogether a disqualification. Both Spangled and Pencilled Hamburghs are subject to white face, as well as the Blacks, but in the latter it is much more common; white patches under the eye and near the ear-lobe frequently making their appearance, especially when a bird is out of condition or old. Indeed, it is rare to find a two or three-year-old cock *quite* free from it. But this is another fault, which may be obviated entirely by careful breeding.

"The legs should be of a *dark* leaden hue; they get lighter with age, but in a young bird a light-coloured leg is to be avoided.

"A Black Hamburgh, if a good specimen, is a most attractive bird. The bright crimson comb and face, the snowy ear-lobe, and the lustrous green of the plumage, form altogether a *tout ensemble* such as no one can see without admiring. Indeed, I have often been surprised to hear lady visitors, when looking through my yards, express greater admiration for the Blacks than the Silver-spangled Hamburghs; these latter being, to my mind, the most striking of all fowls at first glance.

"In a Black Hamburgh cock the breast, back, shoulders, and tail should be of a rich green, the brighter the better; the wing-coverts (which form the bar of the Spangled Hamburghs), exceedingly brilliant; and the outer web of the secondaries—*i.e.*, the whole of the lower part of the closed wing—almost as bright; the lesser tail-coverts are also very rich in colour.

"There are two distinct shades of colour. Some strains are of a deep *blue* green, almost a steel-blue—these have green tails; other strains are of a lighter green—these have bronze green tails. I much prefer these latter myself; but anyway, the purer the green, and the less admixture of purple or any other tint, the better. Many birds are more or less pencilled with bluish purple, or, as the Lancashire fanciers call it, '*mazarine*;' hens chiefly on the back, and cocks on the flights and tail. I believe it is sometimes caused by weakness or ill-health at the time of the growth of the feather, as I have known birds, which as chickens were quite free from this defect, show it after a late or protracted moult; but more often it is hereditary, and I should not care to breed from a bird which showed it to any extent. The hackle and saddle should be deep black, the longest feathers having the centre or main part bright green, and the outer edges or 'hackly' part of the feather (so to speak) black; the thighs and under parts also black. I think I ought to add that it is rare to find a cock as richly coloured as I have described, very few showing much colour on the breast or hackle. The hen should be of a bright glossy green throughout, especially on the wings and back. Mr. Tegetmeier speaks of spangling being visible; all I can say is that I have examined birds in all lights, and I have never been able to see on a *good* bird anything of the sort.* The *whole* of the feather, except the fluffy part, should be uniform in colour; the green cannot be too *pure*, and the less of blue, or purple, or plum, or mazarine, the better.

"As regards breeding for exhibition, it is one great advantage of this variety that prize birds of both sexes can be bred from the same parents; at the same time, wherever there is convenience for doing so, it is advantageous to have the breeding-stock divided into two or more separate yards. You have more strings to your bow; you are more independent of your neighbours as regards fresh blood; and moreover, it is very difficult to find a cock in which are united all the qualifications for breeding both sexes.

"In the selection of breeding-stock the faults to which Black Hamburghs are most liable should be kept in mind. These are ill-shapen combs, white faces, legginess, red hackles and saddles. I believe all black fowls are subject to this last defect, but Black Hamburghs are so especially, and the richer colour the bird, the more likely the red is to show itself. Now at an exhibition, colour does not seem to be looked for by the judges, in cocks especially, so much as the other points; not nearly so much in my opinion as it ought to be, though, of course, *cæteris paribus*, colour will carry the day; therefore, for cockerel breeding, parents should be chosen with *perfect* combs, good red faces free from white, round white ear-lobes free from red, hackle and saddle *perfectly* free from red, short legs, broad back and chest (narrow-bodied birds too often being squirrel-tailed, which is most objectionable, and generally hereditary). Want of colour is not of so much consequence.

"On the other hand, for breeding pullets, a cock must be sought for with as many of the above qualifications *as can be found* united with very brilliant colour. As I have already said, it is difficult to find a very rich cock quite free from red; but colour you *must* have for breeding pullets, and I would much rather choose for the purpose a red-hackled cock, if good in other respects, than a dull-coloured one. I have often bred beautiful lustrous pullets from hens with very little colour, when mated with a bright cock; but never from a dull-coloured cock, however lustrous the hens with him might be.

"I am sorry that I cannot give any statistics as to the laying powers of Black Hamburghs; I have never kept any written records. But there can be no question as to their excellence in this respect. I have given away many to neighbours and friends, and they have almost all expressed astonishment at the numbers of eggs produced by them. The pullet which is so faithfully represented by Mr. Ludlow in his excellent picture, began to lay in the beginning of November. She was exhibited several times during the winter, which of course stopped her laying for the time, but she commenced again as soon as she was left in peace, and continued laying until the end of the following November, never having shown any inclination to sit. Most poultry-books say that Hamburghs *never* sit; but according to my experience that is quite a mistake, at all events as regards fowls which have unrestricted liberty. I have kept Spangled, Pencilled, and Black Hamburghs, the purest strains of each, and every year I have had more or less hens of each variety broody. Some individuals, like the pullet mentioned above, never attempt to sit; others will be broody two or three times in a season. I have not often allowed them to sit, wanting

---

* Mr. Tegetmeier's mistake evidently arose from examining birds bred by crossing Spangled Hamburghs with Spanish. Many so-called Black Hamburghs are thus produced, and we have even seen such breeding *recommended* and described in poultry journals as the correct procedure. Such cross-bred birds will show the iridescent green spangle Mr. Tegetmeier describes, but which can never be distinguished on good birds of a pure strain.

them for other purposes, but whenever I have done so I have found them quite as steady, and quite as good mothers, as the regular sitting breeds, though this again is contrary to the generally received opinion.

"As a rule all my fowls enjoy perfect liberty, but during the breeding season I am obliged to keep some confined, and these I have found to lay well, remain healthy, and apparently quite contented, in yards about sixty feet long and ten feet wide.

"There is no trouble in preparing Black Hamburghs for exhibition; they require no washing; the smoke of a town does not spoil their good looks. If they have a good grass-run, and are not injured by over-feeding or over-showing, they are always ready at a minute's notice to be put into their travelling hampers. They are also less subject to roup than other Hamburghs, and when they do bring it home with them from a show, they seem to recover more quickly.

"They are, of course, subject to the same drawback as other Hamburghs; *i.e.*, the cocks are never so good for showing after their first year. The ear-lobe loses its smoothness, and, to some extent, its purity of colour. Mr. Beldon seems to have the knack of bringing out old birds in better condition than any one else can do; and I fancy it may be due to his birds being kept more under cover, and therefore less exposed to be scorched by the summer sun.

"In conclusion, I will only say that in my judgment there does not exist a more useful, handsome, or profitable breed of fowls. They are undeniably good layers of fair-sized eggs. They are very good upon the table. It is true they are not so large as some breeds, but neither is their appetite; and I believe, if the quantity of food consumed is taken into consideration, they will show a better balance-sheet at the end of the year than some of their bigger brethren."

Besides the standard breeds of Hamburghs and the Redcaps already mentioned, various other kinds of marking have appeared from time to time at particular shows, or have been seen by individuals at different times. Amongst these have been a Silver Hamburgh beautifully *laced*, which has been seen by Miss Watts; but as these birds were described as being very small, and have never appeared since, it seems to us very probable they were either a cross from the Silver-laced Bantam, or perhaps even a large pen of that variety with a full-feathered cock. There would, however, be not much difficulty in producing a real laced Hamburgh if such were desired, by choosing and developing the crescentic marking of the old Yorkshire pheasant, till it was brought up to the required standard, as in the Polish fowl. A Black Hamburgh *laced with white* has also been spoken of, but its authenticity is doubtful, as there is no certain record of such a marking having ever been seen in any variety of fowl whatever. All buff, without any pencilling or spangling at all, has also been seen, but never, we believe, shown. A variety the colour of which was buff pencilled with *white* appeared on several occasions at shows many years since, but was never encouraged, the effect not being pleasing. It was very probably formed by crossing Golden-pencilled with White, as in the case of Chamois Polish fowls. Pure White Hamburghs have not unfrequently been seen, and Mr. Beldon informs us were formerly bred true to feather, but of late years have died out; their points were the same as other Hamburghs, except being pure white all over. They used to be rather wanting in whiteness of deaf-ear, but this could easily have been corrected. They were rather pretty, and could readily be bred at any time by selecting the lightest Silver-pencils. The most worthy of preservation of all the extraneous varieties is, however, the Cuckoo Hamburgh, described in the following notes by Mr. Beldon:—

"I used," he says, "to have a breed of Hamburghs that were very handsome, of a cuckoo colour. They had all the characteristics of Hamburghs except the colour being cuckoo, both cocks and hens being alike. As usual in any variety of this marking, some of the cocks had red in their saddles, while others were free from it. [This is a common fault in all cuckoo or Dominique varieties.] They were capital layers, and in my opinion a very pretty breed, but found no favour at the shows, and therefore have nearly died out. They bred very true to colour, indeed the cuckoo is very easy to breed if you once get it on one side, and by a cross with some other variety the points might have been worked up well." We are inclined to think a cuckoo-colour might meet with more favour now than formerly, some other varieties having made it rather more familiar and popular than in old times.

Silver-pencilled Hamburghs were formerly termed Creoles or Creels in some localities (evidently from the mingling of black and white in the plumage); and an American poultry magazine published in 1873, gave a short description and engraving of a pair of birds under this name. In this case, however, the fowls were evidently bad or ill-bred *Spangled* Hamburghs—*i.e.*, bred larger and coarser by the sacrifice of feather. Such birds, like the Redcaps, are often extraordinary layers.

Mr. Serjeantson has already stated that Black Hamburghs will do well in runs of only moderate size, and we have also known Silver-spangles maintain their high qualities in small yards, provided they were on a dry soil and kept *rigidly clean*, which is absolutely essential to them under such circumstances. The other varieties are only adapted to a grass-run, and lose much of their prolificacy if penned up, being also then subject to roup and other diseases, though when at large the Spangles at least are tolerably hardy fowls. Their great merit of course is as layers, though the meat is excellent so far as it goes. As a layer and table fowl combined the Redcap is probably one of the most profitable that can possibly be, though it lacks those charms of beauty which render the other varieties so attractive. Crossing is not to be recommended in these breeds; but if resorted to, the best is that between a Silver-spangled cock and a Light Brahma hen, which often produces spangled fowls of considerable size and very great beauty. We have sometimes thought a large and handsome spangled fowl might in this way be produced and perpetuated. Chickens of this cross are excellent layers.

Hamburghs require very little to prepare them for exhibition, neither looking well when fattened, nor submitting to the process. They need nothing more than to be in brilliant condition as regards plumage and ear-lobes. Confinement under cover will make a great deal of difference to the ear-lobes of the cock, which become rough and tinged if exposed; and is almost necessary if *old* cocks are to be shown, as the deaf-ears of old birds left at large generally become rough and coarse. The pens should not be less than six or eight feet square, and of ample height, so that the birds may be able to fly up to the perch for exercise. Darkness is not intended, but only to screen them from the wind and sun, which will rapidly improve their appearance; and all cocks, Mr. Beldon says, should be thus put up for about a fortnight, to get them to look bright, as well as used to the pens. The hens should not be put up, as they cannot stand it, but go back in condition; but those which are to be shown should always run together first, and be first tried in a pen to see if they agree. Hamburghs are not generally a pugnacious breed; but it is a singular fact that of the many cases of injury from fighting during exhibition, and particularly of one hen or pullet being scalped by the other, more have been observed by us in the Hamburgh classes than any others; though it must be remembered that Malays and Game are rarely now shown in pairs at all.

We may remark that the purest strains of Hamburghs will occasionally produce birds with single combs, particularly if the smallest and best combs be bred from. This is evidently a case of reversion to the type of some ancient progenitor of the family, and is no proof whatever of a cross in the strain. Such birds will generally breed the proper type of comb; but, for obvious reasons, they should not be bred from, or the tendency will of course increase.

JUDGING HAMBURGHS.—The points to which a judge has to give attention in judging Hamburghs may be ranged into four great divisions, viz., 1, Marking; 2, Combs; 3, Ear-lobes; 4, Symmetry. All of these are important; and symmetry in particular should by no means be overlooked, since it is one of the chief beauties of all the Hamburgh breeds. Tastes differ, and we know many who consider the Game the type of beauty as regards form; but for our own part we must admit that a perfectly-shaped Hamburgh cock, particularly of the Pencilled breeds, is to our own eye far superior—indeed, a very ideal of beautiful contour. Of the other three divisions, all are more or less frequently made the subjects of deceptive practices, and in *no* breed does "trimming" require such lynx-eyed vigilance on the part of the judge. Ear-lobes are painted white; combs are cut and otherwise maltreated; false tails are fastened in; and in the case of the Spangled breeds we have seen a pretty good basketful of feathers abstracted from one pen of prize birds, which were too heavily spangled, and needed "thinning-out" in order to show sufficient of the ground-colour between. This last fraud is very difficult of detection indeed, and, in fact, almost impossible of absolute proof; and even the fastening in of false sickles cannot be detected in every case without manipulation such as a judge hardly feels justified in using, for fear of injuring the plumage of a really honest fowl. The Silver-pencilled cocks are most frequently subject to this latter fraud, a *perfectly* edged tail being by no means easy to produce, and not unfrequently, when it is, combined with a bad comb, with somewhat too dark body-colour, or other faults. When, therefore, an unusually fine pair of sickles are produced, whether on a good bird or a bad one, they are, by unscrupulous exhibitors, frequently transferred or preserved—as the case may be—"for other uses." The fraud is generally detected in the end, either by the fastening "coming loose," or a little want of freshness, or some other cause arousing the suspicion of the judge; but we have known a man boast when thus discovered that the false tail "had done pretty well after all, for he had won with it eleven times." A very dark, glossy, sharply-edged tail, if found on a cock with very slight bars and little under-colour, or perhaps with no bars at all, should always excite suspicion; and this is perhaps the only hint which can be given; though if the sickles look dull and dead in colour while the rest of the plumage is brilliant and glossy, attentive curiosity may also be occasionally rewarded by unexpected discoveries.

It is, however, in the combs that Hamburghs are subjected to the most extensive manipulation, and some of the practices which have actually been detected are of the most cruel nature. In one notorious case the judges took from a comb two large needles, which had been inserted longitudinally to keep it from falling over. These needles were so rusted in that they were removed with difficulty, and were only detected by the end of one protruding, so that the duration of the torture inflicted on the unfortunate bird cannot be known. Pins, temporarily inserted, have been found in plenty. A very common fault is a hollow or depression in the centre of the comb; and this has been known to be treated by cutting a wedge-shaped piece out of the middle, and stitching the outside portions tightly up together till joined and healed. Stitches put in for one purpose or other are often found, and, we regret to add, are employed far oftener than found, being withdrawn just before sending the bird in—indeed, our impression is that it is to simple forgetfulness of this necessary precaution discovery is often due. It is to be regretted that combs merely *carved* have been for years admitted by the judges to pass with practical impunity. Very recently only have some judges appeared to view the matter differently; but owing to energetic remonstrances on the part of the press, some recent cases have been visited with disqualification, and it may be hoped that comb-cutting will be in future discouraged, if not stopped.

Owing to recent changes, combs have been allotted more points in judging in the following schedules, than in the earlier editions of this work.

SCHEDULE FOR JUDGING HAMBURGHS.

GENERAL CHARACTERISTICS OF COCK.—*Head and Neck*—General appearance of head neat, lively, and smart, rather short than otherwise; beak rather short and small; comb double or rose, wide and square in front, and tapering into a long spike pointing backwards and slightly upwards behind, to be flat on top, and full of "work" or points, and set firmly and upright on the head; deaf-ear flat, and as nearly circular as possible; wattles thin, neat, and rounded; neck rather long, much arched or curved, carried well back, and full of long sweeping hackles, flowing well over shoulders. *Body*—General appearance neat and symmetrical, not tapering to the tail as in the Game fowl; the back a fair moderate length, but appearing rather short from the hackle flowing over it; saddle broad, and amply furnished; wings large, but neatly carried; breast full, round, and carried well forward. *Legs and Feet*—Thighs slender, short, and neat; shanks rather short, thin, and clean; toes very slender, and nicely spread. *Tail*—Very ample, with long and broad sickle-feathers, and plenty of secondary sickles or furnishing-feathers; the sickles much curved and carried high, but not squirrel-fashion or over the back. *Size*—Small, averaging in Spangles about five pounds, but larger development no disadvantage if gained without sacrifice of points. *General Shape*—Light and graceful. *Carriage*—Lively, jaunty, and graceful, but not strutting, as in the Spanish cock.

[NOTE.—The shape of the Pencilled varieties is somewhat more light and slender, with less width of body, and the weight less, than the Spangled and Black.]

GENERAL CHARACTERISTICS OF HEN.—The general characteristics of the cock apply to the hen, with only the usual sexual differences. The same note as to the difference of shape in the Pencilled varieties also applies in her case.

COLOUR OF SILVER-SPANGLED HAMBURGHS.—*In both Sexes*—Beak a dark horn-colour. Comb, face, and wattles, brilliant red. Deaf-ears pure white. Eyes dark hazel. Legs dark leaden blue. *Colour of Cock*—Hackle silvery white, free from yellow, spotted at the bottom as much as possible. Back, shoulder-coverts, and wing-bow white, spotted with black, the spots being long and narrow, owing to the pointed shape of the tips of the feathers. Wing-coverts, each feather white, with a heavy round black spangle at the tip, forming two even bars across the wing. Secondaries white, spangled at the tips, the spangles forming what is called the "stepping" of the wing. Primaries also to be spangled on the ends. Saddle-feathers white, spotted on the ends with black. Breast, under parts, and thighs, white spangled with black, every feather having a rich, round, black spangle at the tip, the larger the better, and so arranged as just barely to show the white between. Tail white on the outside, greyish on the inside, each feather spangled at the tip; the sickles and secondary sickles clear white, with a large distinct spangle at the end of each. *Colour of Hen*—Hackle silvery white, each feather spotted with black on the end, the spots towards the bottom becoming larger and rounder. Back, shoulders, saddle, tail-coverts, breast, under parts, and thighs white, each feather tipped with a large, round, black moon or spangle, not arranged so close as to overlap and appear solid black, but so that the white can just be seen between. Tail-feathers white, with a spangle at the end, which is however seldom perfectly round. Wing-coverts tipped with very large spangles, so as to form two regular bars across the wing; and secondary quills to be white, tipped with a heavy crescentic spangle, so arranged as to appear like "steps" on the end of the wing when it is closed. Primaries also white, tipped with black. The marking to be as uniform all over body as possible; and the hens in a pen must match, not only in this, but in combs and other characteristics.

COLOUR OF GOLDEN-SPANGLED HAMBURGHS.—*In both Sexes*—Beak horn-colour. Comb, face, and wattles brilliant red. Deaf-ears pure white. Eyes red. Legs dark leaden-blue. *Colour of Cock*—Ground-colour rich reddish-golden bay, marked as follows:— Hackle and saddle striped with black, the stripes to be sharp and clear. Back, shoulder-coverts, and wing-bow spotted with black at the tips of the feathers. Wing-coverts heavily spangled with large round spangles, forming two bars across the wing. Secondaries and primaries also spangled on the ends. Breast, under parts, and thighs, heavily spangled with rich round spangles. Tail rich green-black. *Colour of Hen*—A rich reddish-golden bay ground-colour, but in other respects similar to the Silver-spangled, except that the hackle is striped instead of spotted with black, and the tail is black. Except in these respects, in fact, the similarity extends to both sexes.

COLOUR OF SILVER-PENCILLED HAMBURGHS.—*In both Sexes*—Beak horn-colour. Comb, face, and wattles bright scarlet-red. Deaf-ears pure white. Eyes bright red. Legs dark leaden blue. *Colour of Cock*—Hackle pure silvery white. Back, saddle, shoulder-coverts, and wing-bow pure silvery white. Wing-coverts used to be sought heavily-pencilled across upper web, so as to form a slight bar, but a white wing is now preferred by most judges, making the whole body pure white. Secondaries white on lower web, except a strip of black next the quills, and black on inner web except a little grey or white on extreme edge; primaries white on outer web and black on inner web. Breast, under parts, and thighs white, except a few black spots behind the thighs. Tail black in the true feathers; sickles and secondaries rich glossy green-black, with a narrow lacing or edging of pure white all round. *Colour of Hen*—Hackle silvery white. Remainder of plumage, except wing-quills, a pure silvery white ground-colour, each feather pencilled across with black; the pencilling to be as fine or frequent as possible, to go as straight and squarely across as possible, and to be nearly as possible equal in width to the white spaces left between. The secondary quills should also be pencilled, but this has rarely or never been yet attained; and white on the outer webs, with a little undefined marking, is the rule. The tail-feathers should be, and often are, perfectly pencilled. The pencilling to "fall in line," as if continuous lines had been drawn round the bird, as far as possible.

COLOUR OF GOLDEN-PENCILLED HAMBURGHS.—*In both Sexes*—Plumage precisely resembles that of the preceding variety, substituting in the cock a ground-colour of reddish-golden bay, and in the hen a rich gold-colour or orange-gold; the black marking being similar.

COLOUR OF BLACK HAMBURGHS.—*In both Sexes*—Beak black or dark horn-colour. Comb, face, and wattles deep rich red. Deaf-ears brilliant white. Eyes bright red. Legs a deep leaden blue, approaching black. Plumage a deep rich black, brilliantly glossed with metallic green, or sometimes bluish purple—the more gloss, and the greener, the better.

VALUE OF DEFECTS IN JUDGING.

1. SPANGLED HAMBURGHS.

Standard of Perfection.	*Defects to be Deducted.*	
A bird ideally perfect in shape, carriage, colour, markings, &c., and in perfect health and condition, to count in points . . . 100	Bad head and comb	20
	Bad carriage of tail	8
	Stained deaf-ear	10
	Deficiency in bars	12
	,, ,, marking of tail (of Silver-spangled)	8
	Spangling too thick, so as to appear black . .	4
	,, too thin and small, so as to appear spotted	12
	,, irregular, or want of clearness in ground, or any other faults of colour and marking	10
	Want of general symmetry	12
	,, ,, condition	15

DISQUALIFICATIONS.—Single or lopping combs. Hen-feathered cocks. Actually red deaf-ears. Absence of bars on wings. Legs any colour but blue or dark leaden-blue. Wry-tails, or any bodily deformity. Trimmed combs, or any other fraudulent dyeing, dressing, or trimming.

2. PENCILLED HAMBURGHS.

Standard of Perfection.	*Defects to be Deducted.*	
A bird ideally perfect in shape, carriage, colour, markings, &c., and in perfect health and condition, to count in points . . , 100	Bad head and comb	20
	Stained deaf-ear	10
	Tail not properly marked	12
	Hackle marked or spotted	9
	Other faults of colour (chiefly of pencilling in hen)	14
	Want of general symmetry	12
	,, ,, condition	15

DISQUALIFICATIONS.—Single or lopping combs. Hen-feathered cocks. Red deaf-ears. Rusty patch on cock's wing in Silvers, or feathers tipped with white in Gold. Legs any colour but blue or leaden-blue. Wry-tails, or any other deformity. Trimmed combs, or any other fraudulent dyeing, dressing, or trimming.

3. BLACK HAMBURGHS.

Standard of Perfection.	*Defects to be Deducted.*	
A bird ideally perfect in shape, carriage, colour, &c., and in perfect health and condition, to count in points 100	Bad head and comb	20
	Stained deaf-ear	12
	Too long ditto	6
	White face (signs of)	8
	Want of "colour" or gloss	12
	Too long legs and thighs	8
	Squirrel-tail	10
	Want of general symmetry	12
	,, ,, condition	18

DISQUALIFICATIONS.—Single or lopping combs. Red deaf-ears. Red feathers. Legs any colour but blue or dark leaden-blue. Decided white face. Wry-tails, or any other deformity. Trimmed combs, or any other fraudulent dyeing, dressing, or trimming.

www.ingramcontent.com/pod-product-compliance
Lightning Source LLC
Chambersburg PA
CBHW062206220526
45470CB00009B/2942